Jellybeans

by Charlotte Stadler

Table of Contents

What Was in Our Teacher's Bag?

Our teacher came into school today holding a big bag. We all wanted to know what was in it.

He told us that his sister owns a candy store and wants to sell jellybeans. He poured his bag into a big jar. There were a lot of jellybeans.

How Many Jellybeans Are in Each Bag?

He said that his sister wants to sell the jellybeans in small bags.

She wants to put only 10 jellybeans in each bag.

When she scoops up the jellybeans, sometimes she gets more than 10.

She wants to know how many bags she can make if she scoops up more than 10 jellybeans.

Our teacher said that we were going to make a table to show how many bags of jellybeans his sister can make with each scoop.

How Did We Make Our Table?

First, our teacher gave us cups
to scoop up the jellybeans.

My friend scooped up 14 jellybeans.

She put 10 jellybeans into one bag.
She told our teacher that 14 would fill one bag with some left over.

How many jellybeans did she have left over?

She had 4 left over.

Our teacher put her answer on our table.

Someone else scooped up 21 jellybeans.
He put 10 of the jellybeans
into one bag and counted the ones
he had left over.

Do you think he could fill
a second bag with jellybeans?

He could. He put 10 jellybeans into a second bag. Now he had only 1 jellybean left over.

Our teacher put his answer
on our table.

What Did Our Table Look Like?

Our teacher put all of our answers on our table.

This is what our table looked like now.

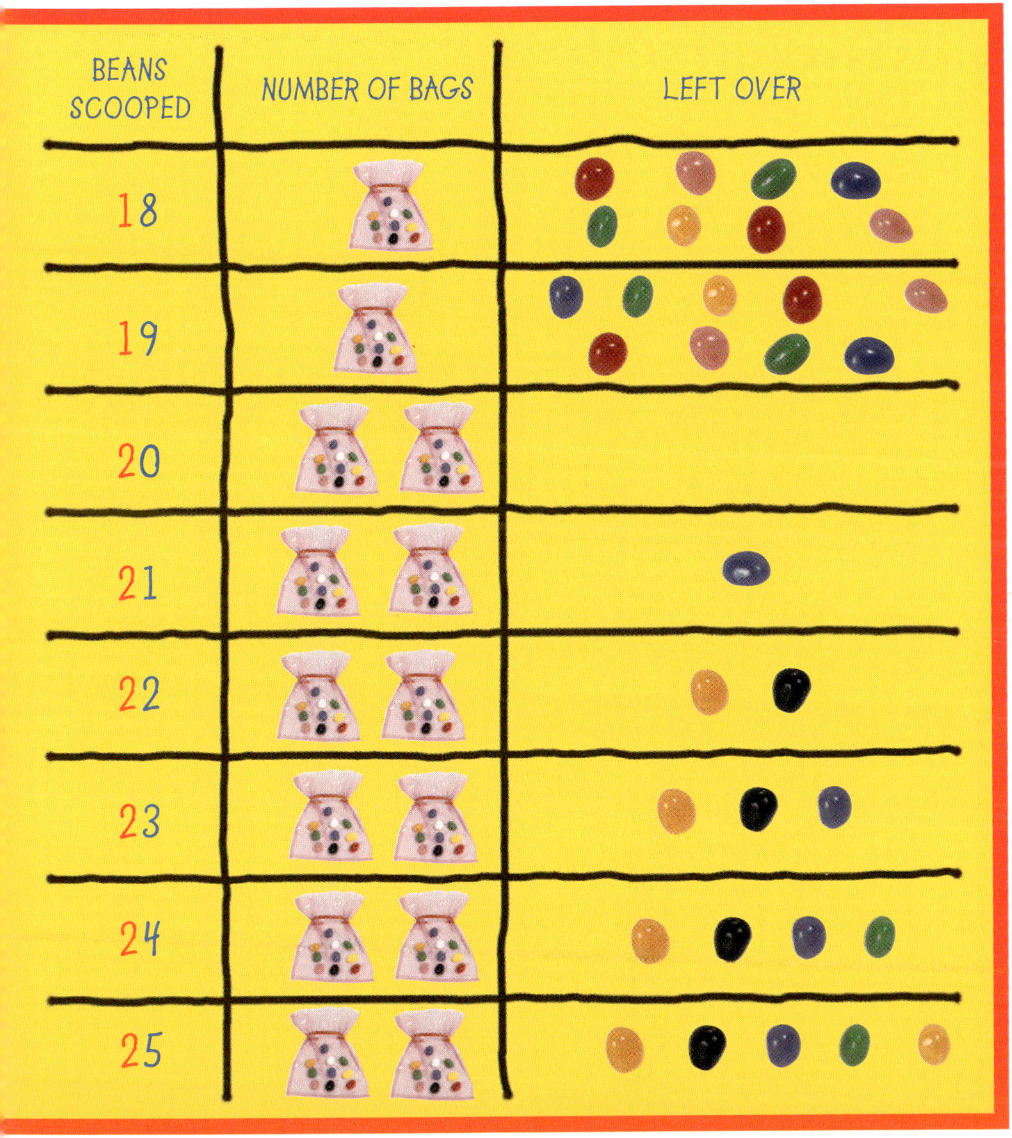

What Did Our Table Look Like When We Used Numbers?

Then we made a table
using only numbers.

The table showed that our red numbers
stand for groups of tens.

The table showed that our blue numbers
stand for ones.

How many bags could you fill
if you had 45 jellybeans? How many
jellybeans would be left over?

NUMBER OF JELLYBEANS	NUMBER OF BAGS	LEFT OVER
10	1	0
11	1	1
12	1	2
13	1	3
14	1	4
15	1	5
16	1	6
17	1	7
18	1	8
19	1	9
20	2	0
21	2	1
22	2	2
23	2	3
24	2	4
25	2	5

Glossary

jellybeans: small, colorful candies that are soft inside like jelly.

patterns: something that is the same over and over.

scoops: lifts up, as with a spoon.

table: a picture that shows information in rows and columns.

Index